BLOOD

Ancestry to Destiny

Tina Ketch

BLOOD
Ancestry to Destiny

ISBN: 979-8-9893695-8-4
eBook: 979-8-9893695-9-1

Table of Contents

Dedication i

Forward iii

The Bloodline of Truth. 1

So, let's begin! 5

Parents' blood type. 9

What does our blood tell us? 19

Compatible blood types. 21

Blood type careers 35

Blood group food requirements. 41

Blood Types: A Global Cultural Perspective 45

Blood types in the US. 47

Blood types and health. 49

The best exercise for blood types 54

Blood types and soulmates 57

Did you know? 59

Why is our blood necessary? 62

Childhood and young adults dealing with trauma. 65

Conclusion 67

Biography 69

Dedication

The concept of dedication holds the profound meaning of wholehearted commitment to a purpose or task. It involves unwavering loyalty, persistence in pursuing a goal, and a willingness to invest substantial time and effort to achieve something significant. I associate this deep sense of dedication with the word "passion." It is my heartfelt aspiration to share with you all that I have dedicated my life to preparing.

I thoroughly appreciate and cherish your willingness to trust the knowledge I impart, your commitment to engaging with my writing, and your openness to passing on the wisdom you have gained from me to others.

Within each of us resides a fervent zeal we call ambition – an intense desire to thrive, accomplish, and explore in this life journey. It is with unwavering faith and trust that we pursue our deepest aspirations. I am profoundly thankful for the support and encouragement you continue to provide by reading my books. Your eagerness to evolve, learn, and embrace the enthusiasm you possess for life profoundly resonates with me. Thank you for sharing that immeasurable passion with me and allowing me to join your extraordinary journey.

Thank you. Tina

Forward

When I am inspired to write a book, I start by drawing inspiration from my surroundings, personal experiences, and various sources of knowledge. I take the time to reflect on what motivates me and carefully consider the specific audience I aim to connect with through the information I'm ready to share.

My desire flows to those who might benefit from the message I intend to convey, and how my words can positively influence their journey is a crucial part of this initial process.

My primary goal is to impart my knowledge and insights to others. Everyone can benefit from and build upon my accumulated wisdom by sharing the lessons I've gathered. I am passionate about facilitating growth and learning through exchanging knowledge and experiences. My most profound passion is giving unconditionally, sharing knowledge, and motivating and empowering others to unlock their inherent abilities and realize their fullest potential. It brings me great joy and fulfillment to recognize that I can make a positive difference in the lives of others by offering my knowledge and support and motivating them to follow their dreams and interests.

I became fascinated by blood groups at a young age after realizing that my mother had Rh-negative blood. This discovery sparked my interest in learning more about the complexities of blood types and their implications. The distinct characteristics intrigued me, leading me to delve deeply into comprehensive research and literature to understand blood groups and their significant importance better.

Moreover, my deep-seated fascination resides in epigenetics's intricate and enthralling realm, where genetic and environmental influences shape our biological and developmental processes. In my writing, I delve into the fascinating complexities of cellular memory and explore the profound interplay between genetics and environmental factors. Through the lens of my research and experience, I have authored three thought-provoking books that delve into the profound interrelations between these fundamental aspects of human life. From a young age, I've been fascinated by the idea of blood types and their wide-ranging impacts on our everyday lives. This deep-seated fascination has spurred me to further explore the importance of blood groups and the profound impact they have on our health and well-being.

The Bloodline of Truth.

Our understanding of complex ideas varies among individuals and is influenced by diverse factors, including individual vibrational frequency, upbringing, education, and environment. These factors collectively contribute to cognitive abilities, surpassing genetic inheritance or academic achievements.

In this book, I will delve into the impact of familial heritage on an individual's capacity to comprehend intricate philosophical concepts and gain profound insights. This investigation will closely analyze how cultural, historical, and ancestral factors shape intellectual and spiritual development. The main objective is to offer an in-depth understanding of cognitive abilities and philosophical insight to make these accessible and meaningful to all individuals.

Building upon our prior research on DNA and genetic memories, our focus will transcend the intricate molecular and cellular domains as we explore intrinsic concepts surpassing life's fundamental building blocks.

Our genetic inheritance from our ancestors significantly shapes our identity. Every facet of our genetic makeup reflects the traits and characteristics passed down through generations, with our physical traits serving as visible markers of our genetic heritage, representing a tangible link to our past. It is vital to acknowledge that inherent factors, rather than being acquired or learned, significantly contribute to shaping our abilities and limitations.

While accomplishments stemming from environmental or educational pursuits are plentiful, inherent limitations within our physical form can pose insurmountable barriers. Genuine progress can only be achieved by understanding and nurturing our innate talents.

Each individual possesses unique capabilities that define their individuality and contribute to the richness and diversity of human experience. We fully appreciate our collective potential by embracing and celebrating these distinctive qualities.

Our blood type is a unique marker of individuality and heritage, coursing through our bodies like sap flowing through a tree and testament to our unique lineage.

Throughout your life, you have grown accustomed to perceiving the world with a limited perspective, akin to looking through a narrow keyhole. However, when you are allowed to broaden your view, you hesitate to embrace the opportunity due to its misalignment with your existing beliefs and preconceptions. I now offer you the chance to expand your vision, enrich your viewpoint, and gain a fresh understanding of what life entails.

Across various cultures, diverse beliefs and interpretations exist about the significance of blood types and whether they hold any defining characteristics beyond their practical use in transfusions. These beliefs vary greatly and are integral to cultural understanding and perception.

Given the profound influence of genetics on our traits and characteristics, it's worth noting that blood type, in conjunction with genetic makeup, DNA/RNA composition, and familial heritage, can serve as a unique marker contributing to the formation of individual personalities. This implies that specific blood types may be associated with certain personality traits and predispositions, highlighting the intricate interplay between genetics and personality development.

In Japanese culture, the belief that a person's blood type influences various aspects of their life is deeply rooted. A person's blood type can significantly impact their temperament and personality traits, affecting their work dynamics and relationships. This belief has led to the common practice of asking individuals about their blood type, which is believed to play a role in everything from matchmaking to job placements. This cultural phenomenon shapes numerous social interactions and can influence important decisions in Japanese society.

In the United States, blood types are determined by the presence or absence of specific antigens on the surface of red blood cells. These antigens, such as the A, B, and RhD antigens, can trigger an immune response if recognized as foreign by the recipient's body. Therefore, it is crucial to carefully match the donor's blood type with the recipient's to avoid adverse reactions. This process, known as blood typing and cross-matching, is essential for ensuring the safety and effectiveness of blood transfusions. Minimizing the risk of an immune response involves healthcare professionals identifying and matching the specific antigens present in the donor's blood with those of the recipient. This in-depth matching process aims to prevent immune reactions that could harm the patient.

Furthermore, it is essential to note that up to this point, there is no definitive evidence or concrete scientific correlations that establish a direct relationship between blood types and profound personality functions or dysfunctions in the United States or Australia. Further research is needed to draw any substantial conclusions in this regard.

So, let's begin!

Blood types were discovered by Karl Landsteiner, a pioneering Austrian physician who worked at the Pathological-Anatomical Institute of the University of Vienna, which is now known as the Medical University of Vienna. In 1901, Landsteiner made a groundbreaking discovery by identifying the ABO blood group system. Subsequently, in 1940, the Rh factor was also determined. The ABO system includes blood types A, B, and O, as well as the Rhesus blood type, which provides for both negative and positive variants. The intricate variations in our blood composition can profoundly influence individual health and determine compatibility with others. It's truly remarkable how such subtle differences can yield such significant effects.

Blood types are determined by specific proteins called antigens present on the surface of red blood cells. There are four main blood types: A, B, AB, and O. Additionally, each can be further categorized as Rh positive or Rh negative based on the presence or absence of the Rh protein. This combination of antigens and Rh factor makes each blood type unique.

Knowing your blood type is essential for a variety of reasons. For instance, it is crucial for medical procedures such as blood transfusions, where matching the blood type of the donor and recipient is necessary to prevent adverse reactions. Additionally, knowing your blood type is essential for pregnant individuals, as any incompatibility between the blood types of the mother and the fetus can lead to health complications. Furthermore, individuals considering organ donation or transplantation need to know their blood type to ensure compatibility with potential recipients.

Understanding your blood type can play a critical role in medical care and make a difference in emergencies. Therefore, individuals must be aware of their blood type and its implications.

Determining your blood type is typically done through a simple blood test, although it's only a routine procedure if necessary. Home blood type testing kits are also available and can provide quick results. Still, it's important to note that these kits are for informational purposes only and may not be as accurate as tests done in a medical setting.

A comprehensive book can provide essential information about blood groups, their significance in medical contexts, compatibility with other blood types, potential health implications, and dietary considerations associated with different blood groups.

1. **Classification**: Blood types are classified into four main groups: A, B, AB, and O. This classification is based on the presence or absence of specific antigens (A and B) on the surface of red blood cells. Blood type A has A antigens, blood type B has B antigens, blood type AB has both A and B antigens, and blood type O has neither A nor B antigens. This classification is essential for blood transfusions and organ transplants to ensure compatibility between donors and recipients.

2. **Compatibility**: Understanding your blood type is incredibly important, especially when receiving transfusions. It's crucial to match the blood type correctly to ensure the safety and effectiveness of the transfusion process. In emergencies, having this knowledge can be life-saving, as it allows medical professionals to swiftly provide the correct type of blood.

3. **Rh Factor**: The Rh factor, also known as the Rhesus factor, is an antigen found on the surface of red blood cells. It can be either positive or negative, which further refines blood types. If a person has Rh-positive blood, their red blood cells contain the Rh antigen, while Rh-negative blood lacks this antigen. When it comes to blood transfusions, it's essential to match the Rh factor of the donor blood with the recipient's blood. Rh-negative blood is generally given to Rh-negative patients to avoid any adverse reactions. However, Rh-positive or Rh-negative blood may be given to Rh-positive patients, as they can usually tolerate either type.

People with type O negative blood are considered universal donors, meaning their blood can be given to individuals of any type. Conversely, individuals with type AB-positive AB blood are universal acceptors and can receive blood from any other kind. Interestingly, only approximately seven percent of the world's population has Rh-negative blood.

The prevalence of different blood types varies across the world. In China and India, a significant proportion of the population has blood type B+, while in many European countries, there is a relatively equal distribution between O+ and A+ blood types. However, it's worth noting that Nordic countries exhibit a slightly higher prevalence of A+ blood type than O+. According to the World Population Review, the distribution of blood types in the U.S. is relatively balanced, with 37.4 percent of the population having type O+, 35.7 percent having type A+, and only 8.5 percent having B+ blood type.

The distribution of blood groups varies significantly from country to country, as there is distinct diversity in the genetic makeup of individual families. This diversity is attributed to genetic reasons.

Understanding your own and your partner's blood types enables you to predict the potential blood type of your baby by considering the various combinations of alleles. This information can be valuable in understanding the possible health risks or complications associated with specific blood type combinations.

Parents' blood type.

When the mother and father have type O blood, the baby will always inherit blood group O. This is due to the genetic inheritance of blood types, where the O blood type is recessive, meaning it will always be expressed when inherited from both parents. Therefore, we can confidently and with 100% certainty predict that the baby will have blood group O in this specific scenario.

Two parents with blood group A.

If both parents have blood group A, their offspring could inherit two alleles: an A and an O or two Os. As a result, the children could have blood group A or blood group O. This inheritance pattern would also be observed if both parents had blood group B.

Two parents with blood group B.

When both parents have blood type B, their child can inherit the B allele from one parent and the O allele from the other parent, resulting in blood type B for the child. Alternatively, if both parents pass on an O allele, the child will have blood type O. This is because the B allele is dominant over the O allele, so if the O alleles are inherited from both parents, the child will have blood type O.

Two parents with blood group A.B.

When both parents with different blood types have a child, there is a 25% probability that the baby will inherit blood type A, a 50% probability of the baby having blood type AB, and a 25% chance of the baby having blood type B. This inheritance pattern follows the principles of Mendelian genetics, where the combinations of parental blood types determine the potential outcomes for the child's blood type.

One parent with blood group A and one with blood group O.

When both parents have different blood groups, predicting the baby's blood group becomes more complex due to the increased number of possible combinations. If one parent has blood group A, their genotype could be A.A. or A.O. If their genotype is A.A., the baby will have blood group A. However, if the parent's genotype is A.O., there is a 50% chance that the baby will have blood group A. This is because the A allele is dominant over the O allele, so if one parent has the genotype A.O., the child has an equal chance of inheriting either the A or O allele.

The Rh Factor.

The Rhesus (Rh) Factor is a vital antigen located on the surface of red blood cells. When the Rh protein is present, an individual is classified as Rh-positive, while its absence leads to classification as Rh-negative. For example, if a person has an AB blood group of A.B. and is Rh-positive, their complete blood type would be A.B. positive. Conversely, if an individual has an AB blood group of O and is Rh-negative, their blood type would be O-negative. Combining AB and Rh groups, there are eight potential blood types: A positive, A negative, A.B. positive, A.B. negative, B positive, B negative, O positive, and O negative. This classification system is critical for blood transfusions and during pregnancy, as it determines compatibility between donors and recipients, helping to ensure the safety and effectiveness of medical procedures.

How the Rh Factor Affects Pregnancy.

As an expectant mother with Rh-negative blood, it's essential to understand the potential impact on your pregnancy if your baby has Rh-positive blood, a condition known as Rh incompatibility. During pregnancy, the placenta allows the unborn baby's red blood cells to enter your bloodstream, potentially triggering your immune system to create antibodies against the baby's Rh antigens.

While these Rh antibodies typically do not pose a significant risk in your first pregnancy, they can become a concern for subsequent pregnancies. In subsequent pregnancies, these antibodies have the potential to cross the placenta and attack the baby's red blood cells, leading to a severe condition known as hemolytic disease of the newborn, which can cause life-threatening anemia for the baby.

If the baby's father also has Rh-negative blood, the risk of Rh antibodies is eliminated. However, if the baby's father has Rh-positive blood, your healthcare provider will monitor your antibody levels during pregnancy and delivery. In some cases, you may require an injection of immune globulins, such as RhoGAM, to prevent the production of Rh antibodies and protect your baby's red blood cells from attack. It's essential to discuss this with your healthcare provider and follow their recommendations to ensure a safe and healthy pregnancy.

Knowing your, your partner's, and your children's blood types can provide valuable insights into your family's temperament. This information can help tailor the necessary care and knowledge for each family member according to their specific needs based on their blood type.

It is essential to pay attention to your children's needs, considering their age and temperament. Starting early is crucial when it comes to addressing their unique requirements. One important step is determining your child's blood group, as this can help you swiftly address any specific needs they may have rather than relying on guesswork based solely on their age. Understanding your child's blood group can provide valuable information for their health and well-being.

A Blood group children.

Children with blood type A are often described as possessing many positive attributes. They are thought to be intelligent, passionate, sensitive, and cooperative. These individuals value loyalty and patience and seek harmony and peace in their relationships and surroundings. They are known for their meticulous attention to etiquette and societal standards.

According to the blood type personality theory, individuals with blood type A are inclined to adhere to established etiquette and societal norms and are generally averse to breaking them. Instances, where they do diverge from these standards, are believed to have a greater significance for them and are viewed as self-fulfilling prophecies.

Moreover, your child with blood type A is considered a methodical decision-maker, taking their time to carefully consider all aspects before concluding. They are not known for their multitasking ability; instead, they prefer to focus on one task at a time. These individuals have a penchant for organization and invest significant effort in maintaining order and tidiness in their environment. Their meticulous and consistent approach to tasks reflects their earnest and diligent nature.

Children with blood type A may be very stubborn and easily stressed. They may have higher levels of the stress hormone cortisol, which can cause them to be perceived as intense. They do not like conflicts and prefer to be in harmony with everyone. They also want to work in collaboration with others in the community.

B Blood group children.

People with B blood type are believed to possess characteristics that make them more inclined towards creativity. They are known to make decisions swiftly but may struggle with following orders. Once they set their focus on something, they demonstrate unwavering commitment, refusing to give up even in the face of seemingly unattainable goals. Individuals with B blood type often exhibit a strong drive to excel in anything they dedicate themselves to. However, according to blood type personality theory, they may struggle multitasking, neglecting other essential tasks while fixating on a single objective.

Children with B blood type may experience discrimination due to perceived negative traits such as selfishness, stubbornness, and uncooperativeness. Sadly, society often emphasizes these negative aspects, overshadowing the many positive qualities that individuals with this blood type possess. As a result, they may experience feelings of isolation and may withdraw from social interactions more than those with other blood types.

Children with B-positive blood type often display persistence, fixating on objectives that may seem impractical or unattainable. This trait may stem from their tendency to hold on to things and experiences, refusing to let go quickly.

Individuals with the A.B. blood group, also known as the universal recipient, can receive blood from donors with A, B, AB, or O blood types.

AB Blood group children.

The AB blood type is a rare and unique combination of A and B personality types. Children with this blood type often exhibit a fascinating blend of extroverted and introverted traits, making them appear enigmatic and complex to those around them. Renowned for their exceptional empathy and conscientiousness, people with AB blood type consistently demonstrate an ability to consider various perspectives, coupled with outstanding logical and analytical skills. They approach life with a humanistic outlook, valuing compassion and understanding in their interactions.

Children with the AB blood type display these remarkable traits from a young age, showing empathy and thoughtfulness in their interactions with others. They are known for their exceptional logical and analytical abilities, reflecting a humanistic approach to life and valuing compassion and understanding in their relationships with others.

O Blood group children.

Children possessing blood type O positive are often characterized by their confidence, determination, and optimism. They are commonly perceived as natural leaders, renowned for their passion and resilience.

Additionally, they exhibit outgoing, amicable, and friendly traits, rendering them approachable and well-regarded within social environments. Individuals with blood type O positive are considered universal donors as they can donate blood to individuals with positive blood type. Still, they can only receive blood from individuals with O-positive or O-negative blood type. This is due to the presence of specific antigens on red blood cells. Additionally, scientific studies have shown that individuals with blood type O have a lower risk of developing heart diseases such as coronary artery disease and have a reduced likelihood of experiencing memory-related issues compared to individuals with other blood types. This may be attributed to specific genetic and biological factors associated with blood type O.

Children with blood type O negative are often described as confident, self-determined, strong-willed, and ambitious. They are believed to be natural leaders with a competitive spirit and a high degree of focus and discipline. Moreover, they are typically known to be practical, organized, and responsible, often excelling in stressful situations due to their resilience and adaptability. Children with O negative blood type may exhibit assertive behavior within the family dynamic, frequently striving to assume a dominant parental role.

Given our current understanding of how blood groups can impact our children's temperament, you might be interested in exploring the practical implications of this knowledge. It is essential to consider the enduring influence of our ancestors within us and our future generations, as this emphasizes the unique nature of each individual. This indicates that additional attention and empathy may be required to understand their needs better. Once again, coining the phrase, not everyone is the same for a reason.

Understanding our child's unique personality is genuinely awe-inspiring. Each child carries a blend of traits inherited from ancestors and a combination of characteristics from both parents, making them a beautiful blend of the past, present, and future. Investing in their blood group adds another layer of complexity and uniqueness to your little bundle of joy.

The chart provided below is an invaluable resource for gaining insight into the personalities associated with different blood types.

1. **Type A (The Perfectionist)**:

    - **Strengths**: Hardworking, with a strong focus on creativity and attention to detail.

    - **Weaknesses**: Confident, resilient, and able to handle stress effectively.

    - **Compatibility**: Works effectively with individuals who exhibit Type A and Type A.B. personality traits, demonstrating an ability to build strong and productive relationships with both personality types.

    - **Career Opportunities**: Excels in well-organized settings and finds fulfillment in supporting and assisting others.

    - **Health Considerations**: Individuals in this group may be at a higher risk of developing illnesses closely linked to stress.

2. **Type B (The Maverick)**:

    - **Strengths**: Independent, adaptable, and passionate. Value the freedom to make your own choices and the ability to adjust to different situations. Enthusiasm and dedication drive me to pursue my goals wholeheartedly.

    - **Weaknesses**: The individual can be characterized as impulsive, behaving in unpredictable ways, and occasionally placing their own needs ahead of others.

    - **Compatibility**: Has positive and harmonious interactions with individuals with Type B and A.B. personality traits.

    - **Career Opportunities**: Has a penchant for thriving in positions that involve thinking outside the box, taking on innovative projects, or excelling in entrepreneurial ventures.

- **Health Considerations**: Individuals with compromised immune systems may be more susceptible to contracting infections."

3. **Type AB (The Enigma)**:**Strengths**:

- **Strengths**: Demonstrates versatility by effortlessly adapting to various situations and tasks. Exhibits empathy by genuinely understanding and connecting with others' emotions and experiences. Approaches challenges and problems analytically and logically.

- **Weaknesses**: Sometimes, she can be moody, often finding it hard to decide, and occasionally, she can come across as aloof.

- **Compatibility**: Adaptable and can effectively interact with individuals of all blood types.

- **Career Opportunities**: Excels in positions that demand both imaginative thinking and analytical reasoning.

- **Health Considerations**: Individuals with certain risk factors, such as high blood pressure, high cholesterol levels, smoking, and a family history of heart disease, may have an elevated risk of developing heart disease.

4. **Type O (The Warrior)**:

- **Strengths**: Confident, assertive, and practical best describe this individual. They exude self-assurance and are not afraid to take charge when necessary. With a no-nonsense approach, they tackle challenges head-on and are known for their practical solutions to problems.

- **Weaknesses**: Impatient, risk-prone, and sometimes insensitive individuals may tend to act without considering the consequences, seek out and embrace uncertainty, and occasionally overlook the feelings of others.

- **Compatibility**: Has positive and harmonious interactions with people with blood type O and AB.

- **Career Opportunities**: An individual who excels in roles that involve leading teams and taking decisive action.

- **Health Considerations**: People with this condition may be at a higher risk of developing ulcers and experiencing inflammation in their digestive systems.

What does our blood tell us?

Our blood cells possess a fascinating ability known as cellular memory. Cellular memory is the ability to retain information from past experiences and exposures. The concept suggests that our blood cells may retain imprints or information from specific events, illnesses, and vaccinations.

This discovery suggests that blood cells may contain significant information about an individual's health background. As a result, they could offer valuable insights for medical diagnosis and treatment. Therefore, in the interest of taking every possible precaution, if you have undergone a blood transfusion and subsequently test positive for a condition you have never previously experienced, it is advisable to inform your healthcare provider about the transfusion as this information could be relevant to your current health status and medical treatment. However, this idea still needs to be heavily debated and requires extensive scientific investigation to be comprehensively understood. The cells responsible for retaining this memory are the B cells and T cells, which potentially store information about past illnesses and diseases that an individual may have encountered. This intriguing phenomenon opens up new avenues for research and understanding the potential role of blood cells in remembering past encounters. After receiving a blood transfusion, you must be mindful of possible changes in your body. This is particularly relevant if your blood contains cell memory or specific antigens since these factors can send false signals within your body when new blood is introduced. Receiving a donated organ is a remarkable journey that brings about a significant transformation. Following the transplant, individuals often encounter intriguing phenomena such as developing specific food cravings or embracing new hobbies. These experiences are believed to be influenced by the cellular memory of the new organ, which may impact the recipient's preferences and behaviors. Our immune system is excellent at remembering every germ it has ever fought. When these germs try to invade our bodies again, special cells called B-lymphocytes and T-lymphocytes recognize them and quickly work to get rid of them, stopping us from getting sick. This extraordinary ability of our immune system to remember and effectively respond to past threats is essential for keeping us healthy.

The immune system is a complex network involving various components, of which blood cells play a critical role. White blood cells, or leukocytes, are vital to the immune system's function. They continuously circulate throughout the bloodstream, acting as vigilant patrollers, always looking for potential threats. When encountering pathogens, these white blood cells swiftly recognize and bind to them, assisting in the essential "seek and destroy" process.

Moreover, the immune system has a remarkable capability to remember previous encounters with specific microbes. It accomplishes this through memory cells, such as B-lymphocytes and T-lymphocytes. These memory cells allow the immune system to recognize and swiftly respond to familiar invaders if they reappear in the body, providing a quick and effective defense against potential illness. While blood cells do not possess memory, the immune system's ability to retain a record of past infections and effectively respond to them is truly remarkable.

Compatible blood types.

Type A: Compatible with Type A and A.B. partners. They start as best friends and become lovers.

Type B: Matches well with Type B and A.B. individuals. Passionate but occasionally selfish.

Type AB: Adaptable and can get along with all blood types. Talented and composed.

Type O: Good match with Type O and A.B. Confident and assertive.

The emerging concept delves into the potential influence of blood types on our personalities and life paths. It suggests that our unique characteristics, career trajectories, and even romantic fortunes can be linked to our blood type. Proponents argue that this approach may offer a more precise insight into our traits, as blood type is genetically determined and reflects inherited traits from our parents and ancestors. Consequently, the belief is that our blood type could impact our behaviors and romantic destinies. Moreover, there is growing speculation about the existence of psychological and physiological blood type love compatibility. This has led to a rising trend in "blood type dating," wherein individuals seek partners based on compatibility. For those who believe in the interconnectedness of the mind and body, utilizing a love calculator based on blood type compatibility may present a valuable method for determining romantic compatibility with a partner.

A female and A male

In a romantic relationship between two individuals with blood type A, a deep-seated sense of safety and emotional security exists, accompanied by an abundant supply of love and stability. For the woman in this pairing, this combination catalyzes her personal growth, bringing her immense joy and contentment. Both partners in this relationship are blessed with strong intuitive abilities, allowing them to forge a deep and meaningful connection on an emotional and spiritual level. However, they need to be mindful of the potential for disagreements arising from an over-reliance on mutual understanding, particularly when navigating sensitive issues.

A female and B male

The relationship between an A-blood-type woman and a B-blood-type man is characterized by the well-known' opposites attract' principle. Despite their differences, they complement each other very well, finding common ground in their shared spirituality and warm-heartedness. They both exhibit strong passions, but they also face specific challenges. The woman needs to cultivate patience in her interactions with her partner, while the man must learn to be more reliable and consistent for her. They must recognize that their dissimilarities can serve as catalysts for personal and relational growth rather than sources of division.

A female and AB male

Individuals with blood type A and AB often form strong and affectionate connections in their relationships. They usually start as friends and then transition into romantic partners. Type A women are known for their caring and dependable nature, which appeals to the practical and straightforward AB blood type. However, the AB males' tendency towards somewhat unemotional responses may occasionally lead to minor tensions in their relationship. When the type A partner allows the type AB partner to take the lead, the relationship can offer the stability and protection that she desires, which plays to the strengths of the AB male.

A female and 0 male

The compatibility and dynamics between a female with blood type A and a male with blood type O can be complex yet potentially rewarding. Both individuals possess distinctive personalities that draw them towards each other. The woman with blood type A tends to admire the proactive and assertive nature of the man with blood type O, which complements her qualities. In return, the man with blood type O values the logical thinking and thoughtfulness of the woman with blood type A. Their inherent differences often harmonize, with A's calming influence tempering O's ruggedness, while his stability offers her a sense of security. However, any disruption to this delicate equilibrium has the potential to lead to conflict and emotional distress for both individuals.

B female and A male

The compatibility between a B-blood-type woman and an A-blood-type man is often guided by the principle of "opposites attract." Despite their contrasting characteristics, they share common qualities, such as being spiritual and warm-hearted. The A blood type man typically seeks sustained emotional connection with his partner, while the B blood type woman values her personal space and may sometimes exhibit individualistic traits. For this relationship to thrive, the A blood type man must demonstrate patience and understanding, while the B blood type woman should try to share her thoughts and time. This mutual effort can pave the way for a passionate and fulfilling relationship. However, this pairing may encounter frequent conflicts and confrontations if these efforts are not made.

B female and B male

The dynamic between individuals of the same B blood type is marked by a strong sense of independence and self-assuredness, resulting in a relationship characterized as individualistic, profound, and deeply fulfilling. Both naturally social and outgoing individuals keenly understand one another's needs and desires. However, they must navigate their inclination towards autonomy and maintain separate identities to thrive as partners and effectively collaborate in business endeavors. While their initial self-centered approach may lead to conflicts, the eventual expression of their romantic and emotional sides can help them overcome these challenges and deepen their connection.

B female and AB male

In the context of blood type compatibility, the relationship between a female with B blood type and a male with AB blood type starts with passion and ease. However, as time goes on, they find that understanding and innovation are crucial to the success of their relationship. The type B individual is particularly drawn to adaptability and, at times, finds excitement in the unpredictable nature of their AB partner. In contrast, the AB individual appreciates the friendly and outgoing nature of their type B significant other. As time progresses, both partners may fall into routines and become too comfortable, leading to monotony and boredom. This could lead to disagreements and regret if not addressed proactively.

B female and 0 male

The two individuals described as idealists in the text are a female with blood type B and a male with blood type O. Their collaboration is highly effective, and their relationship is both passionate and stable. The woman with blood type B tends to draw a lot of attention to her talents, and the man with blood type O is exceptionally skilled at flattering her and maintaining her upbeat mood. As a result, her creativity thrives in his presence, and she deeply appreciates his supportive role in her life despite not always expressing it overtly. It is important to note that the man with blood type O requires attention and may occasionally experience emotional outbursts. In these moments, the woman with blood type B should be prepared to offer her support and understanding.

AB female and A male.

The secure and loving relationship between a female with AB blood type and a male with A blood type is characterized by a progression from friendship to romance. Blood type A individuals' caring and dependable nature leaves a positive impression on the pragmatic and simplicity-loving AB individuals. However, the unsentimental reactions of AB females can sometimes pose challenges in their relationship. The AB female's strong-willed nature may lead her to take charge of their dynamic. If she fails to understand her partner, it can inadvertently undermine his feelings, potentially leading to arguments.

AB female and B male

The woman with AB blood type and the man with B blood type initially establish a passionate and effortless connection. However, as time progresses, the key to their success lies in their ability to understand each other and introduce innovation into their relationship. The B blood type man is drawn to the AB blood-type female's adaptable and occasionally unpredictable nature, particularly appreciating her social nature. As time goes by, they may find themselves settling into a routine, which can lead to feelings of boredom. Without addressing this issue, they may be in frequent arguments and experiencing regret.

AB female and AB male.

The two individuals with the blood type "AB" are characterized by a deep sense of friendship and a strong foundation of mutual respect. Both individuals demonstrate high intuition, allowing for a profound and meaningful connection. However, they must exercise caution and not overly depend on their mutual understanding, as disagreements can arise when either party encounters sensitive issues. This underscores the need for open and respectful communication to navigate potential conflicts and maintain the strength of their relationship.

AB female and O male.

The relationship dynamics between a woman with blood type AB and a man with blood type O are fascinating. Despite their inherent differences, they have found a way to complement each other exceptionally well. Their contrasting personalities seem to adhere to the "opposites attract" adage, as they both possess a mix of spirituality and warmth in their hearts. The stability and seriousness of the man with blood type O are naturally drawn to the enigmatic and unpredictable nature of the AB woman. They are two sides of the same coin, with their disparities bringing them closer rather than pushing them apart. Both partners must take the time to understand how their individualities enrich their relationship and promote personal growth.

O female, and A male.

The dynamics of a relationship between a woman with blood type A and a man with blood type A are significant. They can sway in either a positive or negative direction for both individuals. Their shared strong-willed nature draws them to each other's essential traits. The woman, with blood type A, admires the man's assertiveness with blood type O, while he values her logical thinking. She softens her rough edges in his presence, and he finds the stability he craves in her. However, for the relationship to thrive, she must learn the art of compromise, while he needs to cultivate a greater sense of empathy. Should confrontation or sorrow emerge, the equilibrium may be disrupted.

O female and B male.

The dynamic between the two idealists, a woman with blood type O and a man with blood type B, is characterized by a harmonious balance of passion and stability. The man with blood type B naturally garners attention for his talents, and he finds that his creativity flourishes in the presence of the woman with blood type O, who is adept at uplifting and encouraging him. Although the man may not express it frequently, he profoundly appreciates her positive influence on his life. It is important to note that the woman with blood type O also requires attention, and there are times when this need can become pronounced. During these moments, the man with blood type B should be prepared to offer his unwavering support.

O female and AB male.

The dynamics of a relationship between a woman with an O blood type and a man with an AB blood type can best be described by the age-old adage, "opposites attract." Despite their inherent differences, they intricately complement each other due to their shared spiritual and warm-hearted nature. The stability and earnestness of the O blood type align with the adaptability and unpredictability of the AB blood type. Both individuals must take the time to understand that rather than causing a drift, their differences can catalyze personal growth and mutual understanding.

O female and O male.

The relationship dynamics between two individuals with O blood type are intriguing. In this scenario, the O woman typically assumes a leadership role and effectively manages the household affairs. Her O blood type male partner understands this tendency and gives her the trust and space she requires to thrive. Both partners possess strong intuition, fostering a deep and meaningful connection. However, it is essential to note that they should be wary of relying too heavily on mutual understanding, as sensitive issues may lead to occasional arguments within the relationship.

A male and A female.

The strong bond between these two individuals, with blood type A, is a secure and comforting foundation filled with genuine love and unwavering stability. This extraordinary combination nurtures the woman's personal growth and overall fulfillment. Their shared sharp intuition deepens and strengthens their connection, creating a profound and lasting bond. However, they need to approach delicate issues with care and not solely rely on their mutual understanding, as this can lead to potential disagreements.

A male and B female

The dynamics of a relationship between a B-blood-type woman and an A-blood-type man can be characterized by the well-known saying "opposites attract." Despite their differences, they share common personality traits, such as being spiritual and warm-hearted. The blood-type man seeks a deep and enduring emotional bond with his partner. In contrast, the B blood type woman values independence and may sometimes exhibit individualistic tendencies. For this relationship to thrive, the man must demonstrate patience and understanding, while the woman should strive to communicate her thoughts and actively invest time in the relationship. By making mutual efforts, they can cultivate a passionate and fulfilling connection. However, if these efforts are unsuccessful, frequent arguments and conflicts may mark the relationship.

A male and AB female

A romantic relationship between a woman with the AB blood type and a man with the A blood type can be secure and loving, characterized by a progression from friendship to deep emotional connection. The A partner's affectionate and reliable nature will resonate well with the AB partner's preference for realism and simplicity. However, potential challenges may arise from the AB partner's pragmatic and sometimes unsentimental responses to situations. Furthermore, the AB partner's inclination to take the lead and assert control in various aspects of the relationship could lead to misunderstandings and conflicts, particularly if the A partner's feelings must be fully understood and appreciated. This could result in significant arguments within their relationship.

B male and A female

The relationship dynamics between a woman with type B blood and a man with type A blood can be seen through the "opposites attract" principle. Despite their inherent differences, they share commonalities in their spiritual and warm-hearted natures. The man, with type A blood, is driven by a desire for enduring emotional connection with his partner. In contrast, the woman with type B blood values her personal space and tends to be independent. By demonstrating patience and understanding, the man can create the emotional support the woman needs while she can strive to express her thoughts and invest more time in the relationship. They can find the potential for a deeply passionate relationship when these efforts are made. However, conflicts and confrontations will frequently arise in this pairing without these adjustments.

B male B female

The connection between these two individuals is characterized by independence and drama, which ultimately leads to a sense of fulfillment. Both individuals, sharing the same blood type, possess highly sociable qualities and a deep understanding of each other. Their bond often transcends beyond being romantic partners and can extend to successful business collaborations. However, they must navigate their strong inclination towards independence and learn to work separately to achieve this. Their mutual self-centeredness may lead to conflicts unless they embrace and express their romantic and emotional sides, which are integral to their personalities.

B male and AB female

The relationship dynamic between a woman with AB blood type and a man with B blood type is characterized by an initially electric and seamless connection that captivates both partners. The woman's flexible and occasionally unpredictable nature intrigues the man with B blood type, while she is drawn to his friendly and outgoing personality. However, as time progresses and the novelty wears off, they may find themselves settling into predictable routines, leading to feelings of boredom. They need to recognize this potential shift and proactively work towards maintaining excitement and understanding in their relationship to avoid discord and remorse.

B male and O female

In romantic relationships, a fascinating dynamic exists between a woman with an O blood type and a man with a B blood type. This dynamic is often described as an intriguing blend of passion and stability. This unique combination suggests that B-blood-type males attract considerable attention due to their perceived talents. Conversely, O blood type females are widely seen as ideal partners who provide admiration and a robust support system, which is crucial in nurturing B's spirit and ultimately fostering his creativity's flourishing.

It is worth emphasizing that B genuinely values the significant presence of O in his life, even if he doesn't always express his appreciation openly. This highlights the importance of O blood type females' supportive and nurturing role in the relationship. Additionally, it's important to note that while B-blood-type men are often the focus of attention, the reading also suggests that O-blood-type females have their own needs for attention. Consequently, it recommends that B blood type men be prepared to offer support during these instances, thus creating a more balanced and fulfilling relationship dynamic.

AB male and A female

The relationship between a woman with blood type A and a man with blood type AB tends to be deep and affectionate, rooted in a strong foundation of friendship and emotional connection. They value being best friends before deepening their connection into a romantic partnership. The loving and reliable nature of the type A woman can deeply impress the pragmatic and simple-loving AB man with her exceptional qualities. However, challenges may arise due to the AB partner's occasional lack of sentimentality. Nevertheless, by allowing the AB partner to take the lead, especially in areas where he excels, the type A partner will find the security and stability she desires.

AB male and B female

In the initial stages of their relationship, the woman with blood type B and the man with blood type AB share a deep and passionate connection that seems to flow effortlessly. However, as their relationship progresses, they realize the need for understanding and innovative approaches to maintain its success. The woman's initial enthusiasm is sparked by the adaptability and occasional unpredictability of the man with blood type AB, while the man appreciates his partner's friendly and outgoing nature. As time passes, they both find themselves falling into a routine, seeking comfort, and ultimately experiencing boredom. Without proactive measures to address this, their relationship could deteriorate, potentially leading to arguments and feelings of regret.

AB male and AB female.

The bond between two individuals who share the AB blood type is built upon a solid foundation of friendship and mutual respect. Both parties possess a heightened level of intuition, allowing them to form a profound and meaningful connection. However, it is essential to be mindful that while their shared understanding is a significant asset, it is crucial to avoid becoming overly reliant. There may be occasions when differing viewpoints arise, particularly when navigating delicate and sensitive matters.

AB male and O female.

The relationship between a woman with blood type O and a man with blood type AB exemplifies the age-old adage that "opposites attract." Despite their inherent differences, they have found a way to complement each other exceptionally well. While their characters may be distinct, they share common ground in their spirituality and warm-hearted natures. The stability and seriousness inherent in individuals with blood type O draw them to the adaptability and sporadic behavior of those with blood type AB. Both parties must recognize that their disparities should not push them apart but assist their personal growth. This mutual understanding will enable them to thrive as a couple.

O male and A female.

The dynamics of a relationship involving a woman with type A blood and a man with type O blood are intricate and multifaceted. This union has the potential to manifest as an optimal partnership or a tumultuous entanglement, exerting profound effects on both individuals. Each partner brings distinctive strengths to the relationship that cultivate their mutual attraction. The woman is drawn to the man's decisiveness and proactive nature, while the man admires the woman's thoughtful and rational approach to life.

Moreover, the woman's calming influence softens the man's rough edges, facilitating harmony within the relationship. Simultaneously, the man provides the stability and security the woman instinctively seeks. However, disruption to this delicate equilibrium can potentially yield conflicts and deep sorrow for both individuals involved.

O male and B female.

A beautiful interplay of strengths and support exists in relationships between B-blood-type females and O-blood-type males. The exceptional talents of B-blood-type females are acknowledged and well-supported by their O-blood-type male partners, who excel at uplifting their spirits. The encouragement provided by O-blood-type males plays a pivotal role in fostering and enhancing the creativity of their B-blood-type female partners. Despite being less vocal about it, the B blood type females deeply value and appreciate the unwavering support and generosity of their O blood type male counterparts. However, it is essential to recognize that while O blood type males are supportive, they also require attention and understanding. In some situations, B-blood-type females must be prepared to reciprocate the support and provide the necessary knowledge to their O-blood-type male partners.

O male and AB female.

The intricate interplay between a woman with an AB blood type and a man with an O blood type serves as a poignant illustration of the "opposites attract" phenomenon within relationships. Despite their inherent differences, they share a profound spiritual depth and warm-hearted nature. The dependability and sincerity demonstrated by the man with an O blood type harmonize with the adaptability and unpredictable conduct of the woman with an AB blood type. Their divergent attributes contribute significantly to their personal development and growth as they strive to fathom and value each other's distinctive qualities.

O male and O female.

The relationship dynamics between individuals with blood type O are often characterized as being romantic and stable. Women with blood type O tend to exhibit strong leadership qualities and excel in managing household affairs. Their male counterparts understand and actively support this characteristic, giving them the trust and space required to thrive. Both partners are recognized for their intuitive nature, which contributes to establishing a deep and enduring bond. However, it's important to note that despite their intuition, they must refrain from relying excessively on mutual understanding, as even the most intuitive couples may encounter disagreements when addressing delicate issues. Hence, maintaining open and transparent communication and mutual respect for perspectives is vital in effectively navigating potential conflicts.

Blood type careers

The best careers for A blood type men.

Finance and Accounting: Individuals with type A blood are often known for their meticulous and detail-oriented approach to tasks. They are highly organized and proactive and thrive in environments that require precision and attention to detail. Their conscientious nature enables them to excel in roles that demand thoroughness and accuracy, making them valuable contributors to any team.

Librarian: Extremely well-structured, organized, and highly dedicated to effectively organizing and managing information.

Novelist: People with blood type A are often thought to be creative and enjoy expressing themselves through writing.

Computer Programming: Logical thinking is essential for problem-solving. It involves following a step-by-step process to conclude. Attention to detail is also crucial, as it ensures that nothing is overlooked and all aspects of a situation are thoroughly considered. Logical thinking and attention to detail are essential skills in many aspects of life, including work and decision-making.

The best career for A blood-type woman.

Finance and Accounting: The B blood type ladies are known for their diligent work ethic and meticulous attention to detail, making them well-suited for overseeing and organizing financial matters. Their conscientious and detail-oriented nature allows them to handle complex financial tasks with precision and reliability.

Librarian: Females who are organized and methodical in their task approach tend to thrive in a library setting. This type of environment is particularly suited for women with B blood type.

Novelist: Feel free to explore and express your creativity through the art of writing. Women with B blood type will let their imaginations run wild and unleash their unique ideas, thoughts, and emotions onto the page. Writing provides a powerful outlet for self-expression and creativity, allowing you to craft beautiful narratives, poems, and stories that reflect your inner world. Embrace the written word as a medium for sharing your voice and making a meaningful impact through creativity.

Computer Programming: Coding requires precise, logical thinking and meticulous attention to detail, making it an ideal career choice for a woman with type B blood.

Graphic Design: Blending innovative thinking with precise attention to detail is a characteristic trait of women with B blood type.

The best career for B blood type men.

Detective: With your inherent curiosity and sharp analytical skills, you can thrive in unraveling complex mysteries and solving perplexing enigmas.

Journalist: Given your insatiable curiosity and keen analytical abilities, you are naturally inclined to delve into intricate puzzles and decipher perplexing riddles.

Artist: Your ability to think outside the box and adjust to different situations makes you well-suited for the challenges of investigative reporting.

Craftsperson: Unleash your creativity by exploring a wide range of artistic expressions.

Psychiatrist: Showing empathy and understanding towards others, offering assistance and support to those in need.

The best career for B blood type woman.

Event Planner: Your ability to adjust to new situations and think creatively allows you to excel at planning and coordinating events.

Entrepreneur: Individuals with Type B personality traits often exhibit a propensity for risk-taking and a willingness to embrace uncertainty, making them well-suited for the challenges and opportunities inherent in entrepreneurship.

Designer (Fashion, Interior, Graphic): Express your artistic flair through design.

Travel Blogger or Tour Guide: Feel free to satisfy your curiosity by delving into various experiences and then sharing those experiences with others.

Social Worker or Counselor: A woman with B blood type is recognized for her empathetic nature, characterized by a profound capacity to comprehend and empathize with the emotions and experiences of others. She is consistently prepared to offer support and aid to those nearby.

Best careers for AB blood type men.

Business Analyst: An AB blood type man who can think critically and analyze complex situations is precious when developing and implementing long-term strategic plans.

Scientist or Researcher: Having an AB blood type can contribute to your success in scientific exploration due to the balance of A and B blood type characteristics. This unique combination may enhance your intelligence and creativity, allowing you to excel in analytical thinking and innovative problem-solving within scientific research and exploration.

Diplomat or Negotiator: Your flexibility and ability to understand and share the feelings of others make you well-suited for roles that involve diplomacy and negotiation.

Psychologist: Understanding human behavior aligns with your unique mix of traits. As individuals, we possess diverse characteristics, experiences, and perspectives that shape how we perceive and interact with the world around us. Therefore, delving into the intricacies of human behavior requires an appreciation for this diversity and a keen awareness of how various traits influence our actions and decisions.

IT Professional: Integrate your analytical thinking and problem-solving skills with a deep understanding of technical concepts and tools.

The best career for an AB blood type woman.

Public Relations Company Manager: Your ability to understand and respond to the needs of others while also being flexible and open to different communication styles enables you to excel in building and maintaining solid relationships.

Negotiator: Utilize your logical thinking and problem-solving abilities to identify mutually beneficial solutions.

Teacher: When you have an AB blood type, you are the perfect lady to generously share your unique knowledge and creativity with others, contributing to the growth and enrichment of those around you.

Attorney: Remember to harness your cognitive abilities and ability to effectively advocate for a cause or issue.

The best career for an O blood type man.

Business Leader: Given your natural leadership abilities, you may find fulfillment and success in executive positions such as Chief Executive Officer (CEO) or other high-level management roles.

Entrepreneur: With your unwavering ambition and relentless determination, you possess the ideal qualities to establish and manage your own successful business.

Sales Professional: Your friendly and outgoing personality is well-suited for roles in sales or marketing, where your ability to connect with people and communicate effectively can make a real impact.

Politician: Utilize your belief in your abilities and ability to express yourself clearly and confidently to create a meaningful and constructive influence.

The best career for an O blood type woman.

Business Leadership: As a woman with an O blood type, you may naturally possess leadership qualities and a strong sense of responsibility. Consider pursuing career paths, such as CEO or high-level management positions, that allow you to leverage these strengths. Your innate ability to take charge and make tough decisions could make you well-suited for these roles.

Entrepreneurship: Your ambition and determination make you ideal for initiating and managing your business venture.

Sales or Marketing: Your extroverted personality and ability to connect with people can be a great asset in roles that involve sales and customer interaction. Your natural charm and communication skills can help you excel in sales-related fields, allowing you to build relationships and effectively persuade potential customers.

Politics or Advocacy: Utilize your self-assuredness and determination to influence and bring about positive change effectively.

Blood group food requirements.

Understanding the relationship between your blood type and dietary choices is a fascinating aspect of personalized nutrition. Proponents of the blood type diet believe certain foods can affect individuals differently based on their blood types. For instance, individuals with type A blood are often encouraged to follow a primarily plant-based diet, emphasizing fruits, vegetables, and grains. In contrast, those with type O blood are typically recommended to consume a diet rich in animal protein and limited in grains.

When considering the potential benefits of aligning your diet with your blood type, it's essential to remember that these recommendations are only sometimes accepted within the medical community. Consulting with a healthcare professional or a registered dietitian is necessary before making significant dietary changes based on your blood type.

Prioritizing your holistic well-being involves recognizing the potential impact of your blood type on your dietary preferences. By making informed food choices tailored to your blood type, you may feel more aligned with your body's needs. Some individuals may find that a diet abundant in fresh produce suits them best, while others may thrive on a diet that includes ample protein and complex carbohydrates.

A blood type.

Fruits: Include a variety of fresh fruits in your daily meals to benefit from their diverse vitamins, minerals, and antioxidants.

Vegetables: Ensure that your diet includes ample leafy greens such as spinach, kale, and lettuce, vibrant and diverse vegetables, and cruciferous vegetables like broccoli, cauliflower, and Brussels sprouts.

Tofu: An excellent source of plant-based protein, this food item offers a rich and diverse array of essential amino acids and nutrients necessary for a balanced and nutritious diet.

Seafood: When choosing fish for your meals, consider selecting varieties such as salmon or mackerel. These types of fish are rich in essential omega-3 fatty acids, which are beneficial for heart health.

Turkey: When selecting meat for consumption, it is recommended to opt for lean poultry, as it is lower in fat content than other cuts of meat.

Whole Grains: Incorporate whole grains such as quinoa, brown rice, and oats into your diet for their nutritional benefits and dietary fiber content.

B blood type.

Meat and Poultry: Explore diverse meats and poultry to enrich your meals. Opt for lean choices such as lamb, rabbit, and venison to add depth and flavor to your dishes.

Seafood: Make sure to include salmon, cod, and trout in your diet due to their numerous health benefits. These fish are excellent sources of protein and omega-3 fatty acids, which are essential for heart health and brain function. Incorporating these fish into your meals can help reduce the risk of heart disease, improve cognitive function, and support overall well-being.

Fruits and Vegetables: When choosing your vegetables, aim for a diverse selection of fresh produce. Incorporate leafy greens like spinach and kale and vibrant options like beets and carrots into your meals. For fruits, opt for a variety that is in season to enjoy a range of flavors and nutritional benefits.

Gluten-Free Grains: When selecting grains, consider incorporating millet, oats, and rice into your diet for variety and nutritional balance.

Dairy Products: Incorporate a moderate amount of low-fat dairy products into your diet.

AB blood type.

Tofu: A fantastic option for getting your protein from plant-based sources.

Seafood: Make sure to incorporate nutritious fish selections like salmon and trout for a delicious and wholesome meal.

Dairy: Regarding incorporating probiotics into your diet, yogurt and kefir are excellent choices. They are fermented dairy products that contain beneficial bacteria for gut health.

Green Vegetables: Include plenty of leafy greens in your diet. They are packed with essential nutrients like vitamins, minerals, and fiber for maintaining good health. Examples of leafy greens include spinach, kale, arugula, and Swiss chard. Incorporating these greens into your meals can help meet your nutritional needs and support your overall well-being.

O blood type.

High-Protein Foods: I favor consuming lean cuts of beef, pork, and poultry, such as skinless chicken and turkey, and seafood, such as fish and shellfish. These protein sources are rich in essential nutrients such as iron, zinc, and omega-3 fatty acids, vital for overall health and well-being.

Fish: Make sure to include fish like salmon in your diet, as it is rich in omega-3 fatty acids, which are beneficial for heart and brain health. Omega-3 fatty acids are known for their anti-inflammatory properties and may help reduce the risk of chronic diseases.

Vegetables: To lose weight, consider incorporating nutrient-dense vegetables such as broccoli, spinach, and kelp. These vegetables are low in calories and high in essential nutrients, making them an excellent choice for supporting weight loss goals.

Fruits: Indulge in an assortment of ripe, juicy fruits to tantalize your taste buds and enjoy a burst of natural flavors and nutrients.

Healthy Fats: Olive oil has numerous health benefits, such as being rich in monounsaturated fats and antioxidants. Furthermore, it adds a delightful and rich flavor to your dishes, enhancing the overall culinary experience.

Blood Types: A Global Cultural Perspective

Blood type significance varies across cultures; some countries have more cultural beliefs associated with blood types than others. Here are a few notable examples:

1. **Japan**: Japanese culture places a significant emphasis on blood types. A person's blood type is believed to influence their personality traits, compatibility with others, and career choices. It is common for people in Japan to ask about someone's blood type as an icebreaker or to gain insight into their character. This belief is so ingrained in the culture that it is often considered in various social and professional interactions.

2. **South Korea**: Like in Japan, in South Korea, there is a belief that a person's blood type is linked to their personality, temperament, and compatibility with others. This belief influences how people perceive and form relationships with one another. For example, individuals with blood type A are often stereotyped as meticulous and perfectionistic, while those with blood type B are viewed as creative and free-spirited. Similarly, people with different blood types may be seen as having better or worse compatibility based on these stereotypes.

3. **Taiwan**: There is a widespread belief in the correlation between blood types and personality traits in Taiwanese culture. This belief is so ingrained in society that it is common for people to discuss blood types in social contexts and even for some employers to consider blood types during job interviews. This practice stems from the belief that certain blood types are better suited for specific roles due to their associated personality traits.

4. **Ecuador**: In Ecuador, there is a high prevalence of individuals with Type O+ blood, accounting for approximately 75% of the population. While not directly intertwined with cultural beliefs, this statistic underscores the significance of comprehending blood types for medical reasons. Understanding the distribution

of blood types within a population is crucial for ensuring effective medical treatment and blood transfusions.

Blood types in the US.

O-positive blood type, also known as O RhD positive (O+), is the most common in the United States, with a prevalence of approximately 38% in the population. It is characterized by antigen D on the surface of red blood cells. This blood type is considered a universal donor for RhD-positive recipients and is essential in blood transfusions and organ transplants. Among the different blood types, it is interesting to note that type O is more prevalent among Latino Americans, with around 53 percent of this ethnic group having this blood type. When comparing different ethnic groups, it's found that approximately 47 percent of African Americans and 37 percent of Caucasians have this specific blood type. According to the latest statistics, A-positive is the second most prevalent blood type in the United States, with a frequency of around 33 percent among Caucasians.

Although state-level data on the distribution of blood type B is not readily available, it's important to highlight that this blood type is relatively less common. Approximately 10% of the population has B-positive blood, while only 2% have B-negative blood. It's important to emphasize the need for blood donations from individuals with type B blood due to the relatively low prevalence of this blood type. Donation centers are critical in consistently raising awareness and engaging eligible type B blood donors to ensure an adequate supply for those in need. AB- blood type is considered the rarest in the United States, as it is found in only about 0.4% of the population.

It's always fascinating to explore the distribution of blood types among different ethnic groups and regions. Understanding this variation is crucial for gaining insights into the genetic diversity of our population. While specific state-level data may not be available, delving into the nuances of blood type variation offers valuable knowledge about our society.

Blood types and health.

Blood type can impact various aspects of health beyond transfusions. While scientific evidence in this area is limited, there are some potential associations worth noting:

1. **Disease Susceptibility**:

   - Research indicates that there may be an association between particular blood types and specific health conditions. For example:

     - Individuals with blood type O may have a lower risk of developing heart disease and certain types of cancer compared to those with other blood types. This is due to specific properties of blood type O that may provide some protective benefits against these health conditions.

     - Type A blood has been associated with a higher risk of developing stomach cancer.

     - People with type AB blood might have a higher tendency to develop blood clots than those with other blood types. It's essential for individuals with type AB blood to be aware of this potential risk and to speak with their healthcare provider about any concerns.

   - The relationships and connections identified may not apply across all situations and warrant additional research and investigation.

2. **Digestive Health**:

   - The Blood Type Diet is a nutritional approach suggesting individuals with different blood types follow specific dietary guidelines to achieve optimal health. For example:

- Type A: Vegetarian diet.
- Type O: High-protein diet.

- Although the concept of this diet is intriguing, more robust scientific evidence is needed to support its efficacy and safety.

3. **Pregnancy and Rh Compatibility**:

- It's essential to consider the Rh factor during pregnancy, which can be either positive or negative. Suppose the mother is Rh-negative and the baby is Rh-positive. In that case, it can lead to a condition known as Rh incompatibility, which can result in hemolytic disease in the newborn. This occurs when the mother's immune system reacts to the baby's Rh-positive blood, destroying the baby's red blood cells. Healthcare providers must monitor and address any Rh incompatibility issues during pregnancy to ensure the baby's health.
- Rh-negative mothers who are carrying Rh-positive babies require special medical care to prevent potential complications.

4. **Immune Responses**:

- Blood type can impact immune reactions. Research suggests that individuals with Type O blood may exhibit more robust immune responses to specific infections than those with other blood types. This could influence an individual's susceptibility to certain diseases and the effectiveness of vaccinations.
- The presence of blood-type antigens in both the donor and the recipient can significantly impact the success of organ transplantation. Matching the blood type antigens of the donor and recipient is crucial in minimizing the risk of

rejection and improving the overall success of the transplant procedure.

Blood type is just one of many factors that can influence individual health. Other significant contributors include genetics and environment.

Specific Rh factors.

The Rh factor, also called the Rhesus factor, is an antigenic protein located on the surface of red blood cells. This protein is essential in the field of medicine and blood transfusions. Here are some intriguing and significant facts about the Rh factor that highlight its relevance and impact in healthcare and blood compatibility.

- **Not a Mutation**: The Rh-negative blood type is a rare blood group that lacks the Rh antigen, also known as the D antigen. This blood type is not a mutation but is derived from the protein substance first discovered in Rhesus monkeys. The mystery surrounding Rh-negative individuals revolves around the question of their origin. The presence of the Rh-negative blood type in specific populations has led to various theories and speculations about its evolutionary and genetic history.

- **Pregnancy Risk**: When a mother and father have different Rh factor blood types, there can be risks during pregnancy. If the baby inherits a different Rh factor from the father, the mother's immune system may produce antibodies against the baby's blood, potentially causing harm to the baby. This is known as Rh incompatibility, and it can lead to a condition called hemolytic disease of the newborn (HDN), which can result in anemia, jaundice, and other serious health issues for the baby. Rh incompatibility is usually managed with medical interventions to protect the baby's health.

- **Rare Blood Type**: A small portion of the global population, approximately 15%, lacks the Rh factor in their blood. The majority, around 85%, is Rh-positive. When it comes to medical procedures such as urgent blood transfusions or organ transplants, individuals with Rh-negative blood may encounter specific challenges due to the limited availability of compatible donors and potential risks associated with incompatibility.

- **Allergies and Food Preferences**: Individuals with Rh-negative blood type may be more susceptible to food allergies. This heightened tendency toward allergies could influence their dietary choices and food preferences.

- **Natural Resistance**: Studies have found that individuals with Rh-negative blood type may possess a natural resistance to certain infectious diseases such as HIV, smallpox, and bubonic plague. Research suggests they may be less susceptible to these diseases than individuals with the Rh factor.

- **Hemolytic Disease**: During pregnancy, Rh-negative women with an Rh-positive unborn baby may develop a condition known as "hemolytic disease of the newborn." This occurs when the mother's immune system produces antibodies that attack the baby's red blood cells, which can lead to severe complications for the baby.

- **Lack of Protective Antigen**: Individuals with Rh-negative blood type lack the Rh antigen on the surface of their red blood cells. This absence of the Rh antigen means that their immune system may be triggered to release antibodies when exposed to Rh-positive blood. This makes Rh-negative individuals vulnerable to specific health conditions and may require special medical attention, particularly during pregnancy and blood transfusions.

- **High Altitude Adaptability**: People with Rh-negative blood have a specific genetic trait that causes their red blood cells to have a higher oxygen-carrying capacity. This characteristic enables them to adapt better to environments with lower oxygen levels, such as high altitudes.

By delving into the intricacies and significance of blood types, we aim to demystify the complexities and inquiries surrounding this subject. Understanding the various blood types allows us to gain valuable insights into our biological makeup and that of others. This knowledge fosters a deeper comprehension of how blood type impacts our interactions, health, and overall well-being.

Combining our understanding of DNA and RNA, genetic memory, epigenetics, and now including blood types, we can form a comprehensive understanding of our existence and the reasons behind it.

Having gained this valuable new information, we are better equipped to navigate our lives effectively. This allows us to make decisions that lead to a more profound sense of fulfillment, tremendous success, and a clearer understanding of our identity and the reasons behind our behavior and choices.

The best exercise for blood types.

Here are some exercise recommendations for individuals with O-negative blood type: Due to their unique physiology, O-negative individuals benefit from exercises that focus on high-intensity interval training, strength training, and cardiovascular workouts. Engaging in activities that promote overall body strength, endurance, and cardiovascular health is essential. Additionally, incorporating swimming, running, and cycling activities can be particularly advantageous for individuals with an O-negative blood type. Always consult with a healthcare professional before beginning any new exercise regimen.

For those individuals with blood type A, it is highly beneficial to incorporate gentle exercises that prioritize relaxation and mental well-being. Activities such as tai chi, aerobics, and yoga are particularly advantageous for maintaining optimal health. These low-impact exercises not only contribute to physical fitness but also promote mental tranquility, which is essential for individuals with blood type A.

Moreover, while gentle exercises are favored, it's important to note that engaging in more vigorous workouts can also be beneficial. However, regardless of the intensity of the training, maintaining mental calmness and composure is crucial for overall well-being. The significance of a serene and composed mind should be emphasized as a critical component of maintaining optimal health for individuals with blood type A.

For people with Blood Type B, it's beneficial to establish a comprehensive exercise regimen to ensure overall balance. Including a diverse range of exercises that promote balance and flexibility is recommended. Here are some specific exercise recommendations tailored for

People with a Type B personality can find a beneficial combination of cardiovascular exercises and mental stimulation to maintain overall wellness and balance. A balance of cardiovascular exercises like running, swimming, or cycling alongside activities that provide cognitive stimulation can help Type B personalities. Engaging in physical and mental activities can support their well-being. For example, consider engaging in tennis, hiking, or golf activities, as they balance physical exertion and cognitive engagement. These activities provide opportunities for moderate physical activity and stimulate the mind through strategizing and skill development.

People with Type B personalities are naturally inclined towards social activities and interaction. They thrive in environments that involve engaging with others and enjoy participating in exercises and activities that allow them to connect with different people. Activities like playing tennis, participating in dance sessions, or joining group exercise classes can be fun and benefit physical and mental well-being.

Pay attention to what your body tells you and discover activities that make you feel genuinely connected and engaged.

Individuals with AB blood type are recommended to follow a balanced exercise routine. This routine should incorporate recommendations suitable for blood groups A and B. This might include cardio exercises like brisk walking or cycling and mind-body exercises like yoga or tai chi. Strength training and flexibility exercises can also benefit individuals with AB blood type.

Jogging or cycling for 40 to 60 minutes at least three times per week is recommended for cardiovascular activities. When you're not jogging or cycling, you can complement these activities with yoga and tai chi to balance your exercise routine and work on flexibility and mindfulness.

Practicing meditation can positively affect the mental well-being of individuals of all blood groups. However, individuals with AB blood type may experience specific benefits from incorporating meditation into their routine.

Blood types and soulmates.

The idea that blood types are linked to soulmates or compatibility is predominantly based on cultural beliefs and popular culture rather than scientific evidence. Research in blood type and personality compatibility is limited, and no concrete scientific basis exists to support the claim. Many cultural beliefs, particularly in East Asian countries, suggest that specific blood type combinations are better suited for relationships. However, it's important to approach such beliefs critically and rely on scientific evidence when considering blood types and relationship compatibility.

1. **Cultural Beliefs**:

    - In certain cultures, most notably in Japan, there is a prevalent belief that a person's blood type can influence their personality traits and compatibility with others. This belief has led to the development of "blood type personality theory," where people are categorized according to their blood type and perceived personality characteristics associated with each type.

    - When people form relationships, they often consider blood type compatibility. This includes not only romantic partnerships but also friendships.

2. **Scientific Perspective**:

    - Based on scientific research, no conclusive evidence supports the idea that blood types play a role in determining soulmate connections.

    - Soulmates are frequently linked with profound emotional and spiritual connections that transcend physical qualities, such as blood type. A profound sense of understanding, empathy, and compatibility often characterizes these connections.

3. **Individual Differences**:

- Each individual possesses a distinct combination of personality traits, values, interests, and communication styles. These factors are crucial in establishing deep and genuine connections with others.

- Soulmates are frequently depicted as individuals who share a deep understanding and connection, complementing each other on a profound level of compatibility and mutual respect.

4. **Love and Compatibility**:

    - Love and compatibility are intricate concepts encompassing many emotional, psychological, and interpersonal dynamics. Deep emotional connections, the sharing of essential life experiences, and a profound mutual understanding between individuals characterize them.

    - Blood type compatibility is unlikely to play a significant role in determining soulmate connections, although it can be an interesting topic for discussion.

Although cultural beliefs and curiosity can influence individuals to consider blood types when forming relationships, genuine soulmate connections surpass such considerations. Finding a soulmate entails much more than simply aligning blood types; it involves discovering someone who deeply connects with your heart and soul, encompassing shared history, soul memory, DNA genetics, and blood types. It's essential to delve deeper when seeking a soulmate, but the inherent feeling of connection often precedes the specific reasons behind it.

Did you know?

In one fluid ounce of blood, an astounding 150 billion red blood cells can be found. These remarkable cells play a critical role in the transportation of oxygen throughout the human body. Just one pint of blood contains around 2.4 trillion red blood cells, underscoring their essential function in the circulatory system. These cells are responsible for carrying oxygen from the lungs to the body's organs and tissues, as well as transporting carbon dioxide back to the lungs for exhalation.

Erythropoiesis, the process of red blood cell production, is a truly fascinating biological phenomenon. Under normal conditions, the bone marrow produces an astonishing 17 million red blood cells per second. This constant production is vital for ensuring the effective transportation of oxygen and supporting proper tissue function throughout the body.

Notably, in times of stress or heightened demand, the production of red blood cells can surge to as high as 119 million cells per second, seven times the typical rate. Red blood cells, also known as erythrocytes, are microscopic, measuring about 7 microns in diameter. To provide some perspective, a micron, denoted as µm, is equivalent to one-millionth of a meter.

It's captivating to observe that when a droplet of blood leaves the heart, it takes only 20 to 60 seconds to circulate through the arteries, effectively delivering oxygen and nutrients to the body's organs and tissues before returning to the heart through the veins. This swift and intricate process underscores the intricate and awe-inspiring nature of the human circulatory system. In one fluid ounce of blood, an astounding 150 billion red blood cells can be found. These remarkable cells play a critical role in the transportation of oxygen throughout the human body. Just one pint of blood contains around 2.4 trillion red blood cells, underscoring their essential function in the circulatory system. These cells are responsible for carrying oxygen from the lungs to the body's organs and tissues, as well as transporting carbon dioxide back to the lungs for exhalation.

Before the influx of foreigners to the Americas, the indigenous people predominantly had blood type O. However, with the arrival of foreigners to the New World, they introduced other blood types, which spread through the intermingling of populations, leading to the emergence of new blood types among the indigenous people. This exchange of blood types is an exciting aspect of the cultural and biological exchange that occurred during this period.

Blood type distribution varies across different populations and regions. Let's explore the distribution of blood types in Europe, mainly focusing on blood group B.

Blood Type Distribution in Europe:

Blood Group B: In Europe, blood group B is more frequently found in Hungarians, Russians, Poles, and other Eastern Europeans. However, it is not found in large numbers among Western Europeans.

Scandinavia and Central Europe: Blood group A is associated with high frequencies in Europe, especially in Scandinavia and Central Europe.

Overall Blood Type Distribution:

The distribution of blood types (ABO and Rh) varies by country. Here are some population averages for specific blood types in a few European countries:

- **Albania**: O+ (34.1%), A+ (31.2%), B+ (14.5%), AB+ (5.2%)
- **Armenia**: A+ (46.3%), B+ (12.0%), AB+ (5.6%)
- **Austria**: A+ (37.0%), B+ (12.0%), AB+ (5.0%)
- **Belgium**: O+ (38.0%), A+ (34.0%), B+ (8.5%)

- **Bosnia and Herzegovina**: A+ (36.0%), B+ (12.0%), AB+ (6.0%)
- **Czech Republic**: O+ (27.0%), A+ (36.0%), B+ (15.0%)
- **Denmark**: A+ (37.0%), B+ (8.0%), AB+ (4.0%)
- **Estonia**: A+ (30.8%), B+ (20.7%), AB+ (6.3%)
- **Finland**: A+ (35.0%), B+ (16.0%), AB+ (7.0%)
- **France**: O+ (36.5%), A+ (38.2%), B+ (7.7%)
- **Georgia**: A+ (34.8%), B+ (21.1%), AB+ (9.0%)

Asian blood groups keep in mind that these percentages are approximate and can vary within specific ethnic groups and regions:

Blood Type Distribution in Asia:

- **O+**: Approximately **39%** of Asians have blood type.
- **A+**: Around **27%** of Asians have blood type A+.
- **B+**: Roughly **25%** of Asians have blood type B.
- **Blood Type Distribution in Italy**:
- The three most common blood types in Italy are:
- **Type O+ (O positive)**: Approximately **45%** of organ donors in Italy have type O blood1.
- **Type A+ (A positive)**: Around **39.5%** of organ donors in Italy have type A blood.
- **Type B+ (B positive)**: The prevalence of blood type B in Italy is not explicitly mentioned but contributes to the overall distribution.

Why is our blood necessary?

The proper functioning of the circulatory system is crucial for sustaining our lives. Regardless of blood type, blood is vital in delivering essential nutrients and oxygen to our cells, removing waste products, and helping regulate body temperature and pH levels.

When considering the impact of Earth's gravitational pull on individuals with different blood types, it is essential to explore the details thoroughly. Venous return, which refers to blood movement from the systemic venous network back toward the heart, is a critical aspect of the circulatory system. It's worth noting that venous return is equal to cardiac output at a steady state due to the operation of the venous and arterial systems in series.

Upon assuming an upright position, the force of gravity affects the vascular volume, causing blood to accumulate in the lower extremities. An easy way to observe this phenomenon is by comparing the size of veins on the tops of the feet while lying down versus standing.

The human body's venous system demonstrates a fascinating physiological phenomenon called venous compliance. Venous compliance refers to the veins' ability to stretch and expand in response to increased blood volume. This high venous compliance allows the veins to accommodate the influx of blood readily. Consequently, a significant portion of the blood volume shifts within the venous system due to gravitational forces.

The force of gravity significantly impacts the distribution of blood pressure in the human body. Even in a typical Earth-gravity environment, when a person is upright, gravity causes the blood pressure to be highest in the lower extremities (legs) and lowest intracerebrally (in the skull).

Our bodies have adapted over time to function efficiently in this environment and have developed intrinsic mechanisms to counterbalance the pressure disparities throughout the body.

Encountering higher magnitudes of gravity, known as G-forces, poses distinct challenges to circulatory regulation. When the body is subjected to larger G-forces, such as during rapid acceleration or deceleration, the difference in blood pressure between the skull and the lower body becomes more pronounced. This can lead to physiological changes and complications in maintaining adequate blood flow to the brain and other vital organs.

In summary, gravity influences the distribution and blood pressure within our bodies, regardless of an individual's blood type. However, it's important to note that specific blood types do not directly impact the gravitational force.

One crucial system affected by gravity is the venous system, critical in returning deoxygenated blood from various organs to the heart. It's worth exploring how different blood types may relate to the functions and characteristics of this essential circulatory system.

Vein Structure:

The structure of veins is intricate and consists of three main layers. The outermost layer, known as the tunica externa, is a robust layer mainly composed of connective tissue, providing essential structural support and protection to the vein.

The middle layer, called tunica media, primarily comprises collagen. The collagen fibers in this layer contribute to the elasticity of the vein walls, enabling them to maintain their shape and withstand changes in blood pressure.

The innermost layer, the tunica intima, comes into direct contact with the blood as it flows through the vein. This layer is lined with endothelial cells and may contain specialized one-way valves. These valves are crucial in preventing the backward flow of blood, ensuring that blood moves efficiently toward the heart and does not pool in the vein. The tunica intima is vital in regulating blood flow and maintaining adequate circulation throughout the body.

In terms of vein types, pulmonary veins have a significant role in the circulatory system by carrying oxygenated blood from the lungs to the heart. Unlike other veins, they transport blood that is rich in oxygen. On the other hand, systemic veins make up the majority of the veins in the body. They are responsible for carrying deoxygenated blood from the body back to the heart for re-oxygenation.

Deep vs. Superficial Veins:

The human body is a complex system with a network of veins that play crucial roles in the circulatory system.

Deep Veins: These veins are located deep within the muscles or alongside the bones. They are equipped with one-way valves essential for regulating blood flow. When the surrounding muscles contract, they exert pressure on these veins, helping to push blood back towards the heart.

Superficial Veins: These veins are in the subcutaneous fatty layer beneath the skin. Like deep veins, they also have one-way valves; however, blood flow in these veins is slower as they do not benefit from direct muscle compression.

Veins are vital for maintaining proper circulation in the body, regardless of an individual's blood type. These blood vessels carry deoxygenated blood back to the heart, allowing for the exchange of nutrients and waste products at the cellular level. Efficient circulation is essential for overall health, as it ensures adequate oxygen and nutrient delivery to the body's tissues and organs while facilitating the removal of carbon dioxide and other waste products. Understanding the function of veins underscores their importance in sustaining optimal physiological function.

Childhood and young adults dealing with trauma.

Blood type A

The experiences of trauma and confronting distant shadows can have a significant impact on the mental and emotional well-being of children and young adults with blood type A. In the face of these challenges, it's crucial to recognize that each person has a distinct way of dealing with them. Tailored support that addresses their needs is essential to help individuals navigate these difficulties effectively. It's important to remember that these traumas may encompass a range of experiences, such as body dysmorphia, which involves an individual's perception of their body, and gender identity, which relates to a person's understanding and expression of their gender.

Blood type B

Childhood traumas involving children with B blood type can be effectively addressed through a comprehensive approach that includes individual counseling sessions as well as group therapy with others who share the same blood type. In addition, specialized treatment tailored to B blood type individuals' unique psychological and emotional needs can be highly beneficial. Ongoing support from compassionate and knowledgeable professionals who deeply understand this blood type's specific challenges and sensitivities is crucial for long-term healing and resilience. These issues may encompass a range of topics, such as victim mentality, which involves individuals perceiving themselves as perpetual victims of circumstances and attributing their failures to external factors.

Blood type AB

Children with AB blood types can be particularly vulnerable to the effects of childhood traumas, which can have a significant impact on their overall well-being and mental health. It's important to recognize and understand how these past experiences have affected them and provide appropriate support and intervention to their blood group when necessary. These children require personalized support tailored to their particular emotional and psychological needs. Understanding and addressing the effects of trauma is essential for promoting their overall development and long-term well-being. Individuals with AB blood type may experience a more significant impact from body shaming, which involves making negative remarks about a person's physical appearance. This harmful behavior can encompass body size, age, hair, clothing, dietary choices, and perceived attractiveness. Whether it's directed towards oneself or others, body shaming can have serious consequences, contributing to mental health challenges like eating disorders, depression, anxiety, and low self-esteem. Support and understanding are essential in this situation.

Blood Type O

Understanding the unique nutritional requirements of children and young adults with O blood types is essential for ensuring their overall health and well-being. These individuals can benefit from a tailored approach to their dietary needs to support their specific blood type. Moreover, it is essential to consider and address any potential health risks associated with this blood type. By providing personalized care and attention to their specific needs, we can significantly contribute to ensuring that they can lead healthy and fulfilling lives, successfully overcoming challenges, and addressing lingering health concerns caused by past shadows. It is important to remember that depression may have a stronger correlation with individuals who have type O blood compared to those with other blood types. It is crucial to provide support and understanding to these individuals, especially children, to comprehend better and address their specific needs.

Conclusion

Understanding the inheritance of blood types is a complex but critical concept in genetics. The ABO alleles, responsible for determining an individual's blood type, are inherited, with one allele coming from the father and the other from the mother. The combination of these alleles inherited from both parents determines a person's blood group. When an individual's blood type aligns with their father's, it indicates a high likelihood of close resemblance in their blood types. This resemblance is due to the genetic material inherited from both parents, resulting in a more similar blood type between the individual and their father. It's essential to note that while a person's blood type may show similarities with their parents, many factors beyond blood type influence their personality traits and other characteristics.

An individual's traits are significantly shaped by a combination of genetic predispositions, environmental influences, upbringing, personal experiences, and cultural background. These factors work together to create a unique framework that defines a person's identity and behavior. The intricate dynamics between these components offer invaluable perspectives into human development and actions. We've got all the knowledge we need right now, and it will be amazing to discover ourselves again through blood types! Enjoy!

Biography

TINA KETCH

Tina embarked on a life-changing journey across Europe for several years. She explored various lifestyles, religions, and beliefs during her travels, immersing herself in the local cultures. This experience allowed her to compare and contrast different religions and cultures, gaining a comparative cultural perspective she had never experienced before.

Tina is an accomplished individual who has dedicated her life to helping others. She is a prolific writer, having authored over 20 books on various subjects such as religious studies, spirituality, metaphysics, and quantum physics. Her books are not just informative but also thought-provoking, and they inspire readers to explore their spirituality and embrace their beliefs.

Tina has a vast knowledge of human behavior, and her insights into the human psyche are invaluable to those seeking guidance. She has helped countless people overcome personal struggles, navigate challenging situations, and find meaning and purpose.

Tina's teachings are based on the idea that people should live their beliefs and practice what they preach. She believes that spirituality and religious dogma are not separate but relatively identical. She encourages people to embrace and live by their beliefs, which is the key to a fulfilling life.

Tina's multifaceted personality makes her an exceptional mentor. She is warm, compassionate, empathetic, and funny;

her ability to connect with others is remarkable. She is an excellent communicator; her lectures are engaging, informative, and thought-provoking. Her teachings are not just informative but also transformative, and they inspire people to make positive changes in their lives.

Additional titles available by Tina Ketch.

www.ingramcontent.com/pod-product-compliance
Lightning Source LLC
LaVergne TN
LVHW010941110826
845149LV00013B/2710

* 9 7 9 8 9 8 9 3 6 9 5 8 4 *